AI: GOOD OR BAD?

NANCY DICKMANN

BROWN BEAR BOOKS

Published by Brown Bear Books Ltd
4877 N. Circulo Bujia, Tucson, AZ 85718
USA
and
Studio G14, Regent Studios, 1 Thane Villas,
London N7 7PH, UK

ISBN 978-1-83572-042-4 (library binding)
ISBN 978-1-83572-048-6 (paperback)
ISBN 978-1-83572-054-7 (ebook)

Library of Congress Cataloging-in-Publication Data available on request

Text: Nancy Dickmann
Consultant: Matthew Lugg
Design and Illustrations: Square and Circus
Design Manager: Keith Davis
Children's Publisher: Anne O'Daly

Picture Credits
The photographs in this book are used by permission and through the courtesy of:
Cover: Freepik.com: **Interior:** Shutterstock: bepsy 12–13, Chokniti-Studio 20–21, Gorodenkoff 8–9, 18–19, Patryk Kosmider 4–5, Monkey Business Images 10–11, El Nariz 14–15, Prostock Studio 16–17, Chaay Tee 6. Artwork: Freepik.com.
All other artwork and photography
© Brown Bear Books.

Brown Bear Books has made every attempt to contact the copyright holder. If you have any information about omissions please contact: licensing@brownbearbooks.co.uk.

Websites
The website addresses in this book were valid at the time of going to press. However, it is possible that contents or addresses may change following publication of this book. No responsibility for any such changes can be accepted by the author or the publisher. Readers should be supervised when they access the Internet.

Words in **bold** appear in the Glossary on page 23.

Manufactured in the United States of America
CPSIA compliance information: Batch#AG/5666

CONTENTS

WHAT IS AI?

We have used computers for many years. They're part of our daily lives. Computers help us do many jobs. And they're getting better all the time! The first computers only did simple calculations. Today, powerful **programs** let them do much more.

Many people already use AI. It powers map **apps** that show the fastest route.

COMPUTERS THAT THINK

You have a brain that thinks and learns. Scientists are working on computers that can do this too. This is artificial intelligence. We call it AI for short. AI is still pretty new. We're still learning how to build and train AI. We're also finding ways to use it safely and responsibly.

Computer or AI?

Computers and AI are not the same. AI is a kind of computer program. It's more complicated than standard programs.

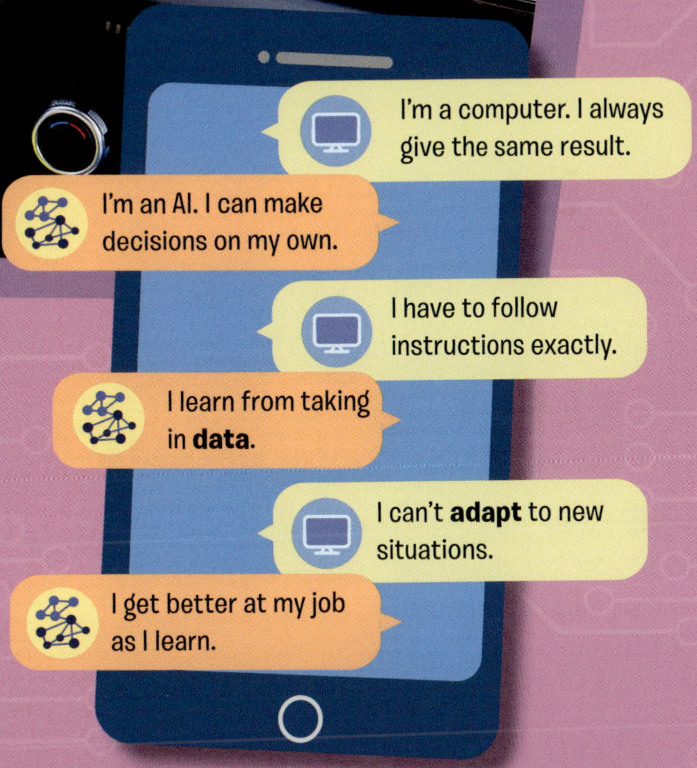

I'm a computer. I always give the same result.

I'm an AI. I can make decisions on my own.

I have to follow instructions exactly.

I learn from taking in **data**.

I can't **adapt** to new situations.

I get better at my job as I learn.

USING AI

We use AI in different areas. One of them is healthcare. Doctors use AI to spot problems on **scans**. AI can also help find new medicines. It can choose the right treatment for each patient. AI can even help hospitals run better.

SAFE TO USE?

There are still many questions about AI. Some people worry that it's too powerful. Others point out that it makes mistakes. How do we build AI? What do we use to train it? Will it take people's jobs? We need to find the answers to these questions.

Dangerous Jobs

Exploring space is dangerous. We can't yet send humans to Mars. We send **robots** instead. AI helps them do their job better. It helps them make decisions on the go.

Cameras have lots of different settings. AI helps use the best ones for each photo.

AI Is Everywhere

AI can help with many different things. Here are a few of the ways we use it.

Online shops suggest products you will like.

Phones unlock by recognizing the user's face.

AI checks crops and tells farmers what they need.

Robot vacuum cleaners use AI to find their way.

AI can keep your home at the right temperatures.

AI helps stores know what to order.

TAKING JOBS

AI does some jobs really well. In fact, it does them better than humans can. It takes a lot of money to build an AI. But using it is almost free. It never gets tired or sick. It never needs a vacation. AI has already taken over some human jobs. It may take over more.

AI also creates many jobs. People are needed to build and train it.

CUSTOMER SERVICE

Customer service agents help customers. They answer phones or chat online. They answer questions. Today, **chatbots** can do this job. They can solve simple problems. Sometimes the question is too hard. Then a human steps in. But we need fewer people doing this job.

AI-powered robots can do the job of a warehouse worker.

Teachers use **empathy** to get the best out of their students.

Who's at Risk?

Some jobs are more likely to be replaced by AI. Others need a human touch. They are safer.

Self-serve checkouts don't need a person to run them.

Construction workers use their hands and skills.

Journalists write news stories. AI can write simple reports.

Managers lead teams and make good decisions.

AI AT SCHOOL

Every student is different. They are good at different things. Teachers want them all to do their best. AI can help with this. It can track how students are doing. It suggests where they need more help. It can make quizzes to test their knowledge.

Some schools allow students to use AI. They want them to learn how to use it safely.

HOMEWORK AND AI

Your teacher might ask you to write a report about frogs. A chatbot could do this task. It will only take a few seconds. But isn't this cheating? The report might get a good grade. But you haven't learned anything. You haven't practiced your writing skills. Some schools have banned students from using AI.

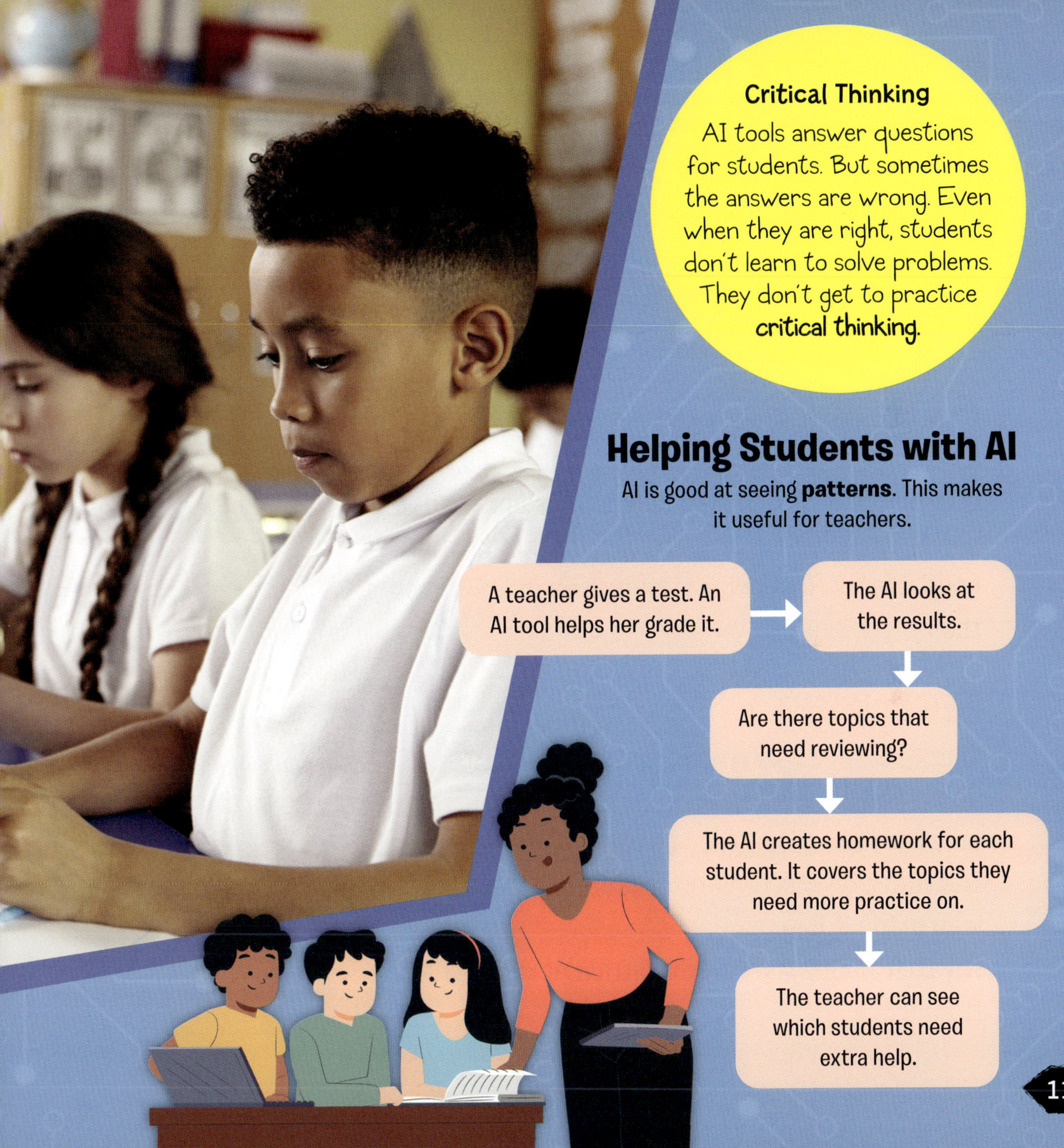

Critical Thinking

AI tools answer questions for students. But sometimes the answers are wrong. Even when they are right, students don't learn to solve problems. They don't get to practice **critical thinking**.

Helping Students with AI

AI is good at seeing **patterns**. This makes it useful for teachers.

A teacher gives a test. An AI tool helps her grade it.

→ The AI looks at the results.

→ Are there topics that need reviewing?

→ The AI creates homework for each student. It covers the topics they need more practice on.

→ The teacher can see which students need extra help.

TRAINING DATA

An AI is able to learn. But it must learn from something! Scientists feed it data. This is called training data. It could be **text** or images. It could even be music. The AI goes through the data. It looks for patterns. This helps it learn. The more it sees, the better it gets.

FREE TO USE?

All this data has to come from somewhere. Some companies take it from the internet. This is called "scraping." They do not pay for it. But this material is often protected. Only the person who made it has the right to sell it. These people say that taking their work to train AI is unfair.

AI learns by looking at examples. But it needs millions of them!

What to Use?

Each AI has a different job. They train on different kinds of data. Here are some common types.

Text from books, articles, or websites

Music tracks

Recordings of people's voices

Charts and tables with numbers

Photos, drawings, or paintings

Movies or TV shows

MAKING MISTAKES

An AI can be a powerful tool. But it can still get things wrong. An AI might identify a photo of a cat as a dog. It might not understand a question or command. AI is trained on lots of data. It knows a lot. But it doesn't really have common sense.

MAKING THINGS UP?

If you ask a chatbot a question, it will give you an answer. It writes something that sounds like it makes sense. That is its job. But it is not trained to tell what is true. That's why sometimes its answers are wrong. You need to check them somewhere else.

Seeing Things

AI helps doctors look at medical scans. It spots some things that they miss. But AI can also miss signs of disease. It can find things that turn out to be nothing.

A chatbot once suggested using glue to make cheese stick to pizza! That would not be safe.

What's the Problem?

AI gets many things right. Here are some reasons why it also gets them wrong.

There wasn't enough training data.

The training data was wrong.

The AI didn't understand the question.

It didn't have the right information.

It can't remember what was discussed earlier.

You asked it to do something it wasn't trained for.

AI BIAS

Bias is when you favor one thing or group over another. You might not realize you're doing it. AI can be biased, too. An AI needs a good range of training data. It must have lots of different examples. If not, the AI will learn to be biased.

WHAT'S THE PROBLEM?

One company tried using AI to decide who to hire. They gave it information. It was about people who had applied before. The AI used that to learn. But most of the data came from men. The AI thought that's what the company wanted. So it picked fewer women to hire. This is an example of AI **bias**.

Some AI has trouble recognizing people of color. It was mainly trained on white faces.

Let users see how the AI makes decisions.

Check the AI's work. Is it fair and balanced?

Fixing Bias

Bias in AI is a problem. But there are ways to avoid it.

Make sure the team making the AI is **diverse**.

Look at the training data. Is it like the real world?

AI HELPING US

AI has its problems. But it can still be a useful tool. People around the world use it every day. Some people use AI for fun, or to learn. Others use it for their jobs. AI does some jobs really fast. It can sometimes do them better than we can.

CHANGING THE WORLD

We can use AI to make our world better. AI is helping scientists find cures for disease. It is helping find ways to fight climate change. We are making new discoveries faster than ever. AI can also help keep us safe. Police use it to solve crimes. It can help us stop **hackers**.

Staying Safe

Floods and storms can be deadly. AI can help **predict** them. It looks for patterns in weather data. This warning saves lives. People can move to a safer area.

AI Helping Out

Here are some ways that AI helps us.

Processes data and does boring jobs

Finds bugs in computer code

Generates text and images quickly

Helps office workers work better

Helps doctors find new medicines

Helps farmers grow better crops

Engineers can use AI. It helps them design better, safer products.

WHAT'S THE VERDICT?

AI is already changing the way we live. It can be really powerful. But it can't solve all our problems. It is just a tool. We have to know how to use it. We must train it in ways that are fair. We also have to check its work. It can't be relied on to always be right.

WORKING TOGETHER

Humans and AIs have different skills. We should give AI jobs that it can do well. Then we keep the jobs that we can do better. People are learning to work with AI. With its help, they can do their jobs better. But they must remember its limits. Then they will get the best from it.

AI may replace some jobs. But it will create others.

Think About It!

AI has its good points and bad points.

It "thinks" faster than a human can.

It can sometimes be biased.

Companies sometimes take people's work to train it.

It can make mistakes.

It often makes our lives easier.

What do you think is the best way to use AI?

QUIZ

How much have you learned about the issues surrounding AI? It's time to test your knowledge!

1. What makes AI different from a standard computer program?

a. it can only follow instructions

b. it learns by taking in data

c. it always gives the same result

2. Which of these jobs is least likely to be replaced by an AI?

a. warehouse worker

b. customer service agent

c. plumber

3. Which word means taking data from the internet to train AI?

a. scraping

b. trolling

c. analyzing

4. Why might a chatbot sometimes give a wrong answer?

a. you didn't pay it enough

b. it trained on bad data

c. it just wants to mess with you

The answers are on page 24.

GLOSSARY

adapt to change in order to become better suited to a new situation

app a computer program designed to do a particular job, usually on a smartphone or tablet

bias favoring one thing or group over another, sometimes without realizing it

chatbot a computer program designed to have conversations with humans

critical thinking the ability to analyze and make a judgment about things you hear or read

data information that is stored or used in a computer, in the form of a series of ones and zeroes

diverse representing a broad range of people from different backgrounds

empathy the ability to understand and share another person's feelings

hacker a person who tries to break into a computer system to steal or cause damage

pattern an arrangement of similarities or trends between different items in a set

predict to make an educated guess about what will happen

program a set of coded instructions for a computer to follow

robot a machine with moving parts that is programmed to do a job

scan an image taken of part of the body to help doctors spot injury or disease

text data in the form of written words

FIND OUT MORE

Books

A Kids Book About AI. Neha Shukla, DK Children, 2025.

Artificial Intelligence. Julie Murray, Abdo Books, 2021.

The Challenges of AI. Lisa Idzikowski, Lerner Publications, 2025.

Websites

www.bbc.co.uk/newsround/49274918

kids.britannica.com/kids/article/artificial-intelligence/390648

www.softwareacademy.co.uk/ai-for-kids/

INDEX

Answers: 1. b; 2. c; 3. a; 4. b